One Duke

Sonja McGiboney

For Theo

"It's okay, Duke!"

Written and published by Sonja McGiboney
ISBN: 978-19643451-1-6

One Duke

1

Woof

Two Dukes

Woof

2

Woof

Three Dukes

Woof

Woof

3

Woof

Four Dukes

Woof

Woof

Woof

4

Woof

Woof

Woof

Five Dukes

Woof

5

Woof

Woof

Woof

Woof

Six Dukes

Woof

Woof

6

Woof

Woof

Woof

Woof

Seven Dukes

Woof

Woof

Woof

7

Woof

Woof

Woof Woof

Eight Dukes

Woof Woof Woof Woof

Woof

8

Woof

Woof

Woof

Woof

Nine Dukes

Woof

Woof

Woof

Woof

9

Woof

Woof

Woof

Woof

Woof

Ten Dukes

Woof Woof

Woof

Woof

Woof

Woof

Woof

Woof

10

Woof

Eleven Dukes

Woof Woof

Woof Woof Woof Woof Woof Woof Woof Woof Woof

11

Woof Woof

Twelve Dukes

Woof Woof Woof Woof

Woof Woof

12

Woof Woof

Woof Woof

Woof Woof

Thirteen Dukes

Woof

Woof Woof

Woof Woof

13

Woof Woof

Woof Woof

Woof Woof

Woof Woof

Fourteen Dukes

Woof Woof Woof Woof

14

Woof Woof

Woof Woof

Woof Woof Woof Woof

Woof Woof Woof

Fifteen Dukes

Woof Woof Woof

Woof Woof

15

Woof Woof

Woof Woof

Woof Woof Woof Woof

That's enough Dukes for now!

If you like this book, please take a moment to review it, or my other books.

https://www.amazon.com/author/sonjamcgiboney

Sign up for my monthly newsletter!

It's full of stories, crafts, guest authors, and more.

https://www.jazzysbooks.com/

Other books that you might like!

Sonja McGiboney creates children's books. Her *Jazzy's Books* series are full of beautiful photographs and are geared to help young children with concepts such as the alphabet, counting, friendship, shapes, and self-worth.

Her works in progress include more picture books, several MG novels, and three adult fiction novels. Her collection of short stories (for adults) is currently with an editor.

She is happy to share that her first chapter book, *A Night in Lacey Manor*, launched in March, 2025.

In her spare time, Sonja teaches piano, is learning to play the violin, enjoys photographing animals, and taking long car trips to photograph scenic roads in the U.S.

Made in the USA
Monee, IL
05 November 2026

67385766R00024